D1462033

The earth is not a mere fragment of a dead history, stratum upon stratum like the leaves of a book, to be studied by geologists and antiquaries chiefly, but living poetry like the leaves of a tree, which precede flowers and fruit,—not a fossil earth, but a living earth . . .

—HENRY DAVID THOREAU

This harvest gift was given to

Patsy

by

Phoebe

on

11 - 2 - 95

Thanks to you & Woods for a lovely visit in your home — and a magnificent evening with "Cinderella"!

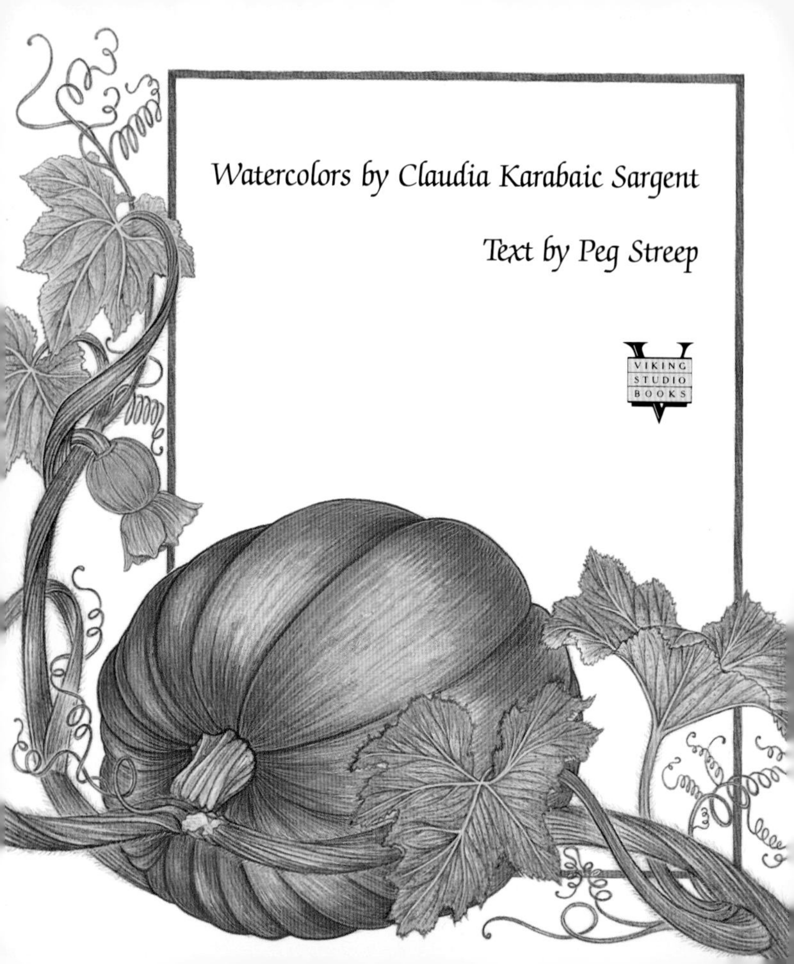
Watercolors by Claudia Karabaic Sargent
Text by Peg Streep
VIKING
STUDIO
BOOKS

A HARVEST Gift

An Illustrated Garden in Miniature

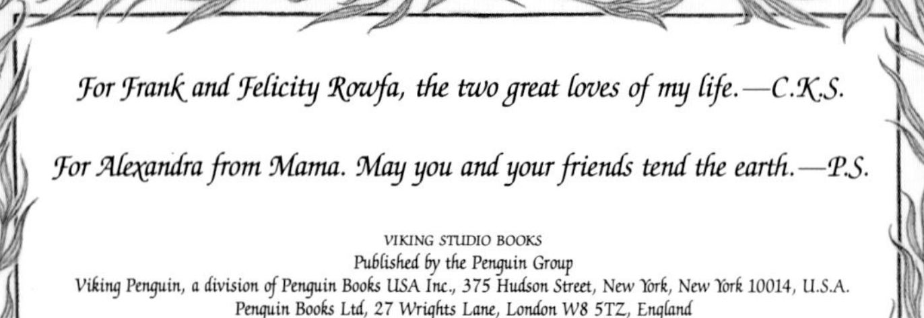

For Frank and Felicity Rowfa, the two great loves of my life.—C.K.S.

For Alexandra from Mama. May you and your friends tend the earth.—P.S.

VIKING STUDIO BOOKS
Published by the Penguin Group
Viking Penguin, a division of Penguin Books USA Inc., 375 Hudson Street, New York, New York 10014, U.S.A.
Penguin Books Ltd, 27 Wrights Lane, London W8 5TZ, England
Penguin Books Australia Ltd, Ringwood, Victoria, Australia
Penguin Books Canada Ltd, 2801 John Street, Markham, Ontario, Canada L3R 1B4
Penguin Books (N.Z.) Ltd, 182–190 Wairau Road, Auckland 10, New Zealand

Penguin Books Ltd, Registered Offices: Harmondsworth, Middlesex, England

First published in 1991 by Viking Penguin, a division of Penguin Books USA Inc.

1 3 5 7 9 10 8 6 4 2

LIBRARY OF CONGRESS CATALOGING IN PUBLICATION DATA
Streep, Peg.
A harvest gift : an illustrated garden in miniature / watercolors
by Claudia Karabaic Sargent ; text by Peg Streep.
p. cm.
ISBN 0-670-83763-6
1. Wild flowers—United States. 2. Fruit—United States. 3. Wild
flowers—Pictorial works. 4. Fruit—Pictorial works. 5. Wild
flowers—Folklore. 6. Fruit—Folklore. 7. Botanical illustration.
8. Wild flowers—Gift books. 9. Fruit—Gift books. I. Sargent,
Claudia Karabaic. II. Title.
QK115.S88 1991 582.13′0973—dc20 91-50151

Printed in Singapore Set in Zapf Chancery Medium

There is, finally, no greater gift than harvest: without it, humanity would find itself huddled on a brown, barren sphere. We have created this book to celebrate the earth's bounty, and the glorious flower- and fruit-laden months from May to November, the months that fulfill the promises of spring. Choosing which of nature's treasures to include has not been easy, but the American wildflower holds a special place in our hearts, and there we start—a trail of pinks and yellows that leads us to the lushness of summer and, finally, to the vivid colors of the autumn harvest. Many of the plants in this book grow without human intervention, although some needed our help to get to these shores; others could not thrive without human efforts. Yet most important of all, as the lore and history we've included will make clear, our life on this planet is not only enriched and beautified by the presence of these plants, it indeed depends on them.

On the pages that discuss more than one fruit or flower, the plants in the borders are identified in the text as they appear starting from the top of the left-hand page, reading counterclockwise.

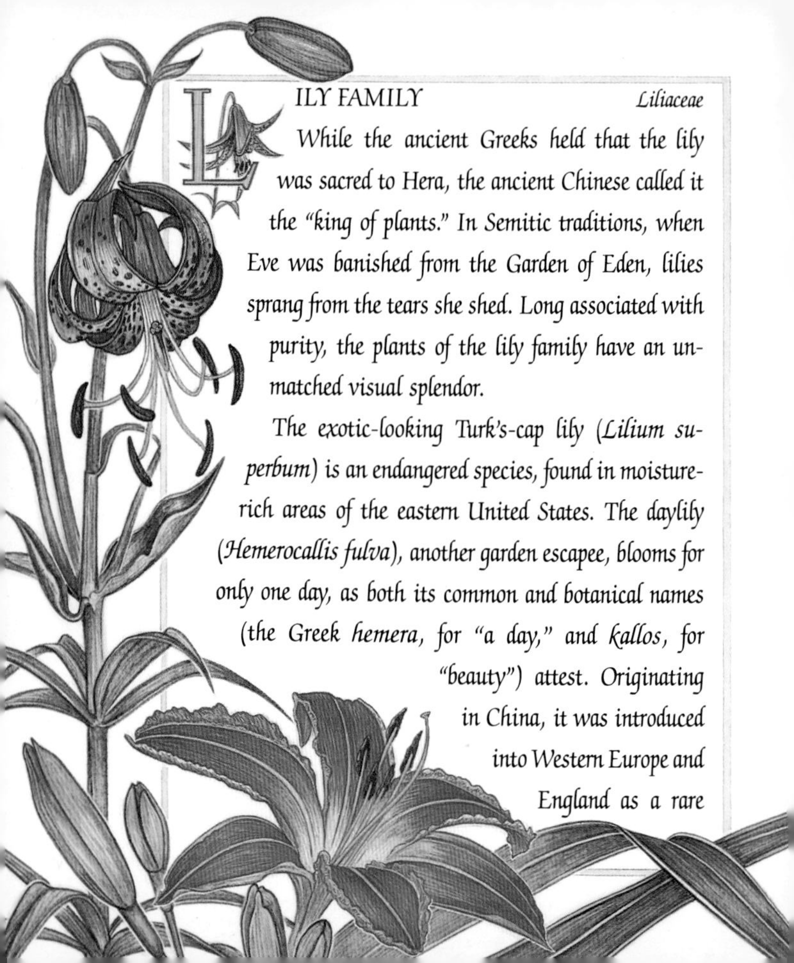

LILY FAMILY *Liliaceae*

While the ancient Greeks held that the lily was sacred to Hera, the ancient Chinese called it the "king of plants." In Semitic traditions, when Eve was banished from the Garden of Eden, lilies sprang from the tears she shed. Long associated with purity, the plants of the lily family have an unmatched visual splendor.

The exotic-looking Turk's-cap lily (*Lilium superbum*) is an endangered species, found in moisture-rich areas of the eastern United States. The daylily (*Hemerocallis fulva*), another garden escapee, blooms for only one day, as both its common and botanical names (the Greek *hemera*, for "a day," and *kallos*, for "beauty") attest. Originating in China, it was introduced into Western Europe and England as a rare

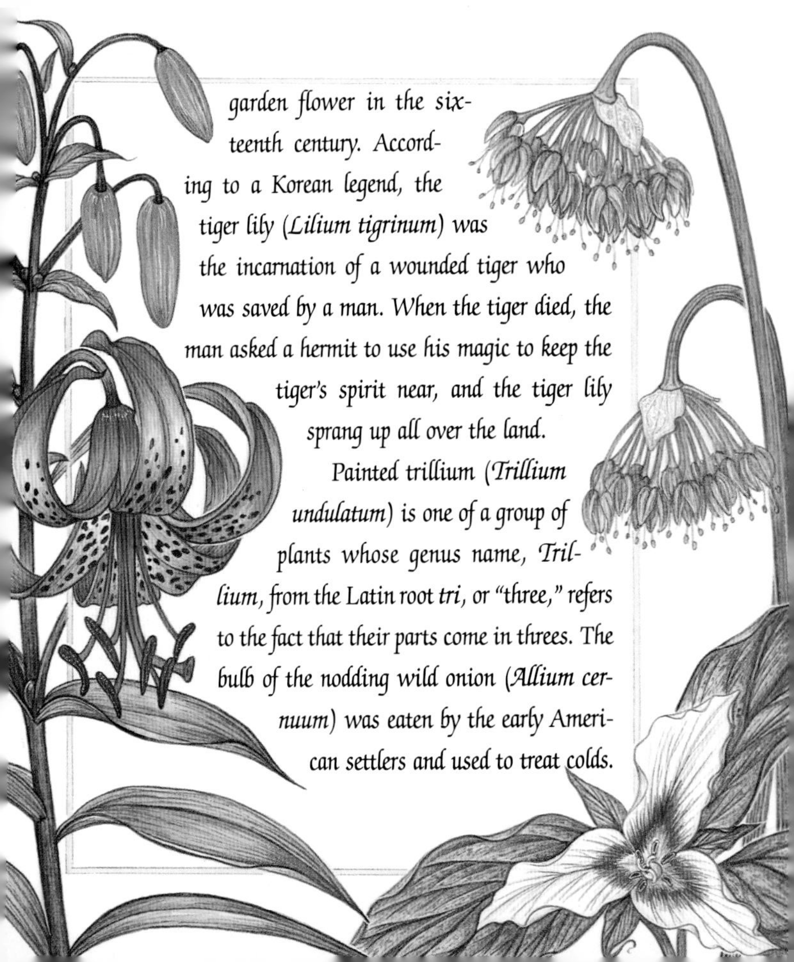

garden flower in the sixteenth century. According to a Korean legend, the tiger lily (*Lilium tigrinum*) was the incarnation of a wounded tiger who was saved by a man. When the tiger died, the man asked a hermit to use his magic to keep the tiger's spirit near, and the tiger lily sprang up all over the land.

Painted trillium (*Trillium undulatum*) is one of a group of plants whose genus name, *Trillium*, from the Latin root *tri*, or "three," refers to the fact that their parts come in threes. The bulb of the nodding wild onion (*Allium cernuum*) was eaten by the early American settlers and used to treat colds.

BUTTERCUP FAMILY *Ranunculaceae*

This family of plants, many of which live near water (hence the genus name *Ranunculus*, or "little frog"), has members with medicinal properties and others that are poisonous. Hepatica (*Hepatica acutiloba*) got its name because the shape of its leaves resembles the liver (*hepatikos* in Greek), and the early herbalists thought it could cure ailments of that organ. Monkshood (so named for the shape of its blossoms), or aconite (*Aconitum uncinatum*), has long been known as a poison. The European species, *Aconitum napellus*, was dedicated by the Greeks to Hecate, goddess of spirits, ghosts, and witchcraft. The larkspur (*Delphinium tricorne*) is also poisonous; its genus name reflects a fancied resemblance of the buds to the dolphin (*delphinus* in Latin). On a slightly brighter note, there is the common or meadow buttercup (*Ranunculus acris*), associated by the Victorians with childhood. Yet buttercups can cause the skin to blister and, in the United States, they were said to cause lunacy. The marsh marigold (*Caltha palustris*) likes moisture, as both its common name and its species name (*palustris* means "of the swamps" in Latin) tell us.

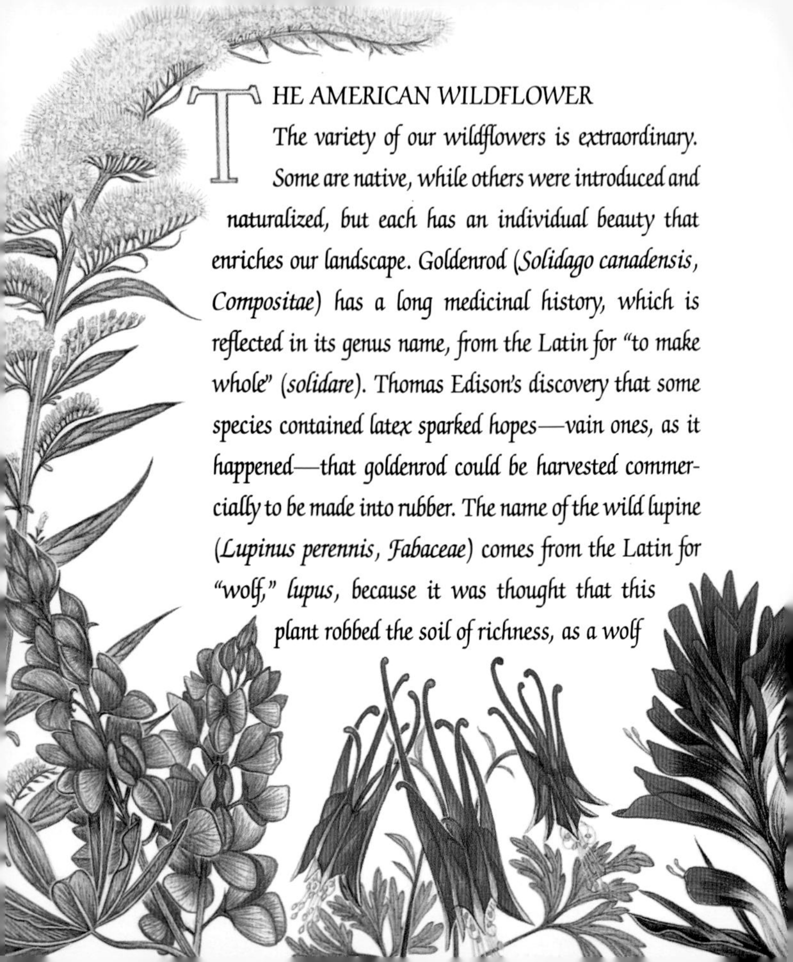

THE AMERICAN WILDFLOWER

The variety of our wildflowers is extraordinary. Some are native, while others were introduced and naturalized, but each has an individual beauty that enriches our landscape. Goldenrod (*Solidago canadensis, Compositae*) has a long medicinal history, which is reflected in its genus name, from the Latin for "to make whole" (*solidare*). Thomas Edison's discovery that some species contained latex sparked hopes—vain ones, as it happened—that goldenrod could be harvested commercially to be made into rubber. The name of the wild lupine (*Lupinus perennis, Fabaceae*) comes from the Latin for "wolf," *lupus*, because it was thought that this plant robbed the soil of richness, as a wolf robs the shepherd of his sheep. (The opposite is true.) The European cousins of this plant were cultivated by the Romans for food (*Lupinus albus*), and by the seventeenth century were touted as medicinal. The wild columbine (*Aquilegia canadensis, Ranunculaceae*) is sometimes called the rockbell, because it grows in rocky places. Both its Latin and its common name have avian associations: *aquilegia* comes from the Latin for "eagle" (*aquila*), since the shape of the flowers' spurs resembles the eagle's talons, and columbine is from the Latin for "dove" (*columba*), since the grouped flowers look like birds in flight. The wild columbine lives up to its linguistic roots, since it is a favorite of hummingbirds.

The Indian paintbrush (*Castilleja coccinea, Scrophulariaceae*) is a member of a genus that includes over two hundred species. Its common name is derived from an American Indian legend: a warrior wanted to paint a sunset but had only his war-paints on hand, so he appealed for help to the Great Spirit, who gave him brushes dipped in the colors of the sunset. Wherever the warrior set down his brushes, the plant henceforth grew. The exquisite musk mallow (*Malva moschata, Malvaceae*) is a naturalized European native known to the ancients, recommended by Pliny, and admired for its medicinal value (it was used to cure inflammations, colds, and hoarseness) and its beauty. The Romans used it to decorate the graves of friends. Bull thistle (*Cirsium vulgare, Compositae*), on the other hand, should only be admired from a discreet distance, since it is the prickliest of our thistles.

Common mullein (*Verbascum thapsus, Scrophulariaceae*) is another European introduction to our landscape, one with a distinguished history. The dried stems of this herb were used as lamp

wicks in centuries past, even though a superstition persisted that witches lit *their* lamps with mullein during incantations.

While yellow flag (*Iris pseudacorus, Iridaceae*) is another cultivated flower gone wild, red clover (*Trifolium pratense, Fabaceae*) was deliberately introduced for use as animal fodder. Early American settlers were happy to discover the cheering native bluets (*Houstonia caerulea, Rubiaceae*), which were also called "Quaker ladies" or "innocence," while the native spotted joe-pye weed (*Eupatorium maculatum, Compositae*) was reputed to have cured the first colonists of a typhus epidemic.

MORNING GLORY *Ipomoea purpurea* *Convolvulaceae*

These pretty trumpets of color please the eye, but they belong to a group of plants that includes the bindweed, a pest that chokes neighboring plants. Like the bindweed, the morning glory grows by twisting around a supporting object, often another plant. Both the genus name, *Ipomoea*, from the Greek for "worm-like," and the family name, from the Latin *convolvere*, "to intertwine," reflect this habit. The morning glory was originally native to Central America; it is said that Cortez sent the first seeds back to Spain after the conquest of the Aztecs. Although early garden writers noticed that the flower opens between morning and noon and withers by evening (a fact which led them to liken it to—and call it—the "life of man"), the common name morning glory is, according to the *Oxford English Dictionary*, an American invention dating from 1836. In the Victorian language of flowers, the plant didn't fare too well: it symbolized affectation.

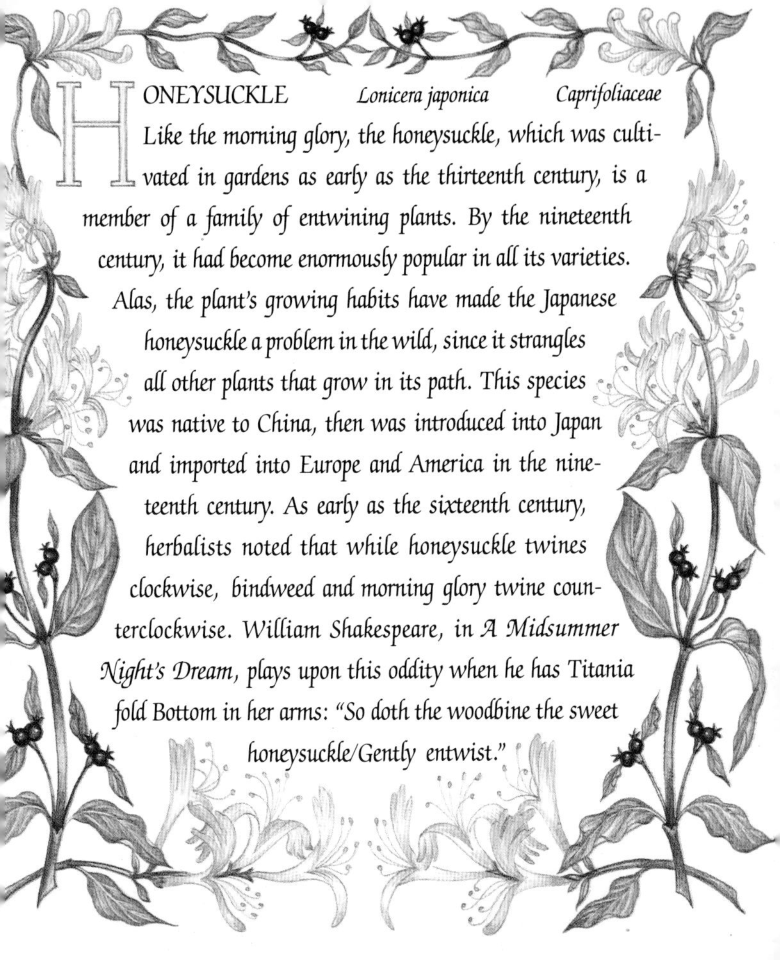

HONEYSUCKLE *Lonicera japonica* *Caprifoliaceae*

Like the morning glory, the honeysuckle, which was cultivated in gardens as early as the thirteenth century, is a member of a family of entwining plants. By the nineteenth century, it had become enormously popular in all its varieties. Alas, the plant's growing habits have made the Japanese honeysuckle a problem in the wild, since it strangles all other plants that grow in its path. This species was native to China, then was introduced into Japan and imported into Europe and America in the nineteenth century. As early as the sixteenth century, herbalists noted that while honeysuckle twines clockwise, bindweed and morning glory twine counterclockwise. William Shakespeare, in *A Midsummer Night's Dream*, plays upon this oddity when he has Titania fold Bottom in her arms: "So doth the woodbine the sweet honeysuckle/Gently entwist."

ROSE FAMILY

Rosaceae

In Shakespeare's words, it is "Summer's ripening breath" that brings us nature in its full glory. That beauty would be diminished without the family we call rose—pears, plums, cherries, apricots, peaches, strawberries, blackberries, and raspberries among them—a cornucopia of riches! First, for some, comes the queen of flowers, the ultimate symbol of love: the rose, or *Rosa rugosa*. But the sweet cherry (*Prunus avium*), too, has long been cultivated by man. In Christian art, the fruit of the cherry symbolizes sweet character, and was sometimes called the "fruit of Paradise." The cherry's delicacy made it a favorite metaphor of poets, particularly for the description of a woman's beauty. The words of the seventeenth-century poet Robert Herrick—which incorporate the cries of a street cherry vendor—sum up a long tradition:

Cherry-Ripe, Ripe, Ripe, I cry,
Full and fair ones; come and buy:
If so be, you ask me where
They do grow? I answer, There,

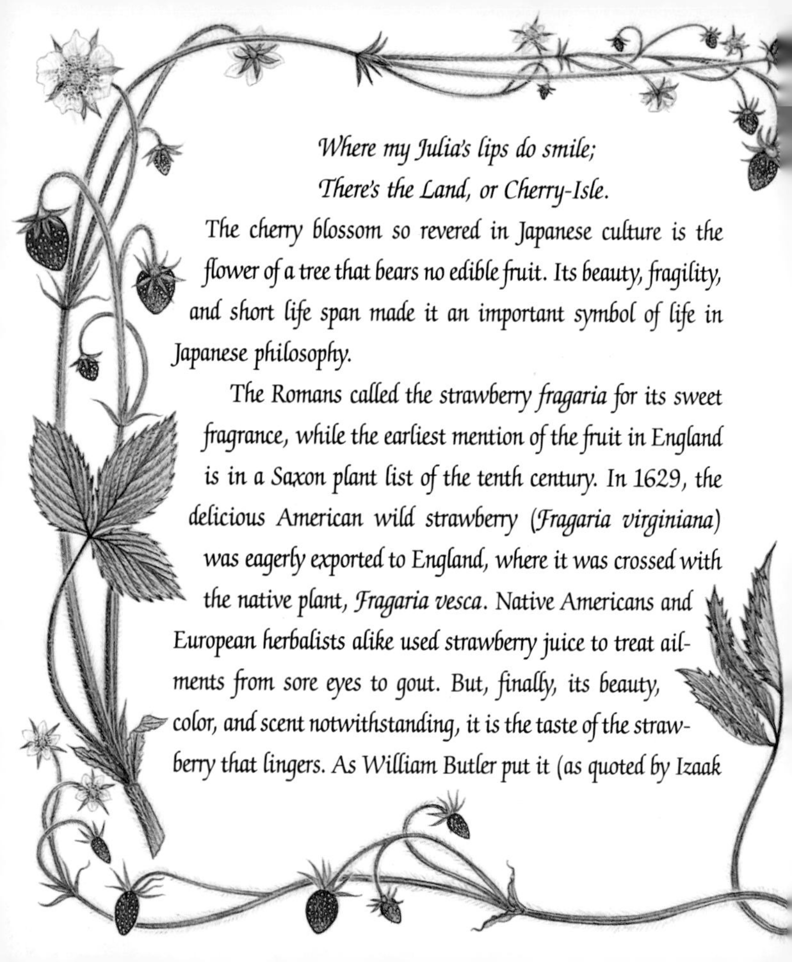

Where my Julia's lips do smile;
There's the Land, or Cherry-Isle.

The cherry blossom so revered in Japanese culture is the flower of a tree that bears no edible fruit. Its beauty, fragility, and short life span made it an important symbol of life in Japanese philosophy.

The Romans called the strawberry *fragaria* for its sweet fragrance, while the earliest mention of the fruit in England is in a Saxon plant list of the tenth century. In 1629, the delicious American wild strawberry (*Fragaria virginiana*) was eagerly exported to England, where it was crossed with the native plant, *Fragaria vesca*. Native Americans and European herbalists alike used strawberry juice to treat ailments from sore eyes to gout. But, finally, its beauty, color, and scent notwithstanding, it is the taste of the strawberry that lingers. As William Butler put it (as quoted by Izaak

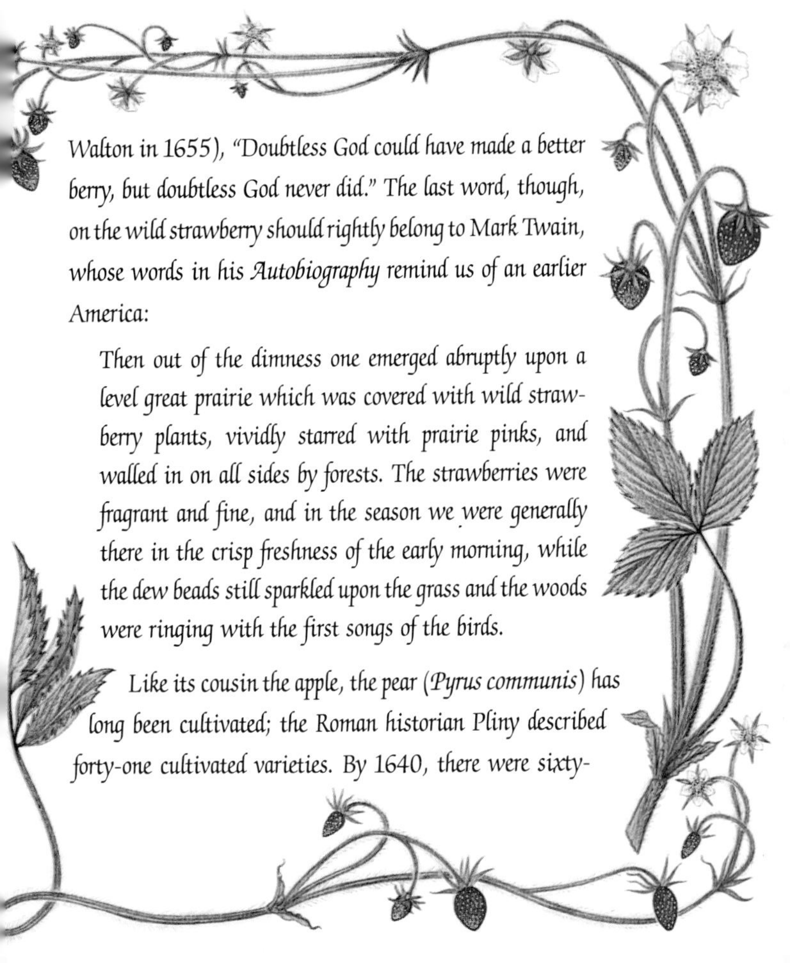

Walton in 1655), "Doubtless God could have made a better berry, but doubtless God never did." The last word, though, on the wild strawberry should rightly belong to Mark Twain, whose words in his *Autobiography* remind us of an earlier America:

> Then out of the dimness one emerged abruptly upon a level great prairie which was covered with wild strawberry plants, vividly starred with prairie pinks, and walled in on all sides by forests. The strawberries were fragrant and fine, and in the season we were generally there in the crisp freshness of the early morning, while the dew beads still sparkled upon the grass and the woods were ringing with the first songs of the birds.

Like its cousin the apple, the pear (*Pyrus communis*) has long been cultivated; the Roman historian Pliny described forty-one cultivated varieties. By 1640, there were sixty-

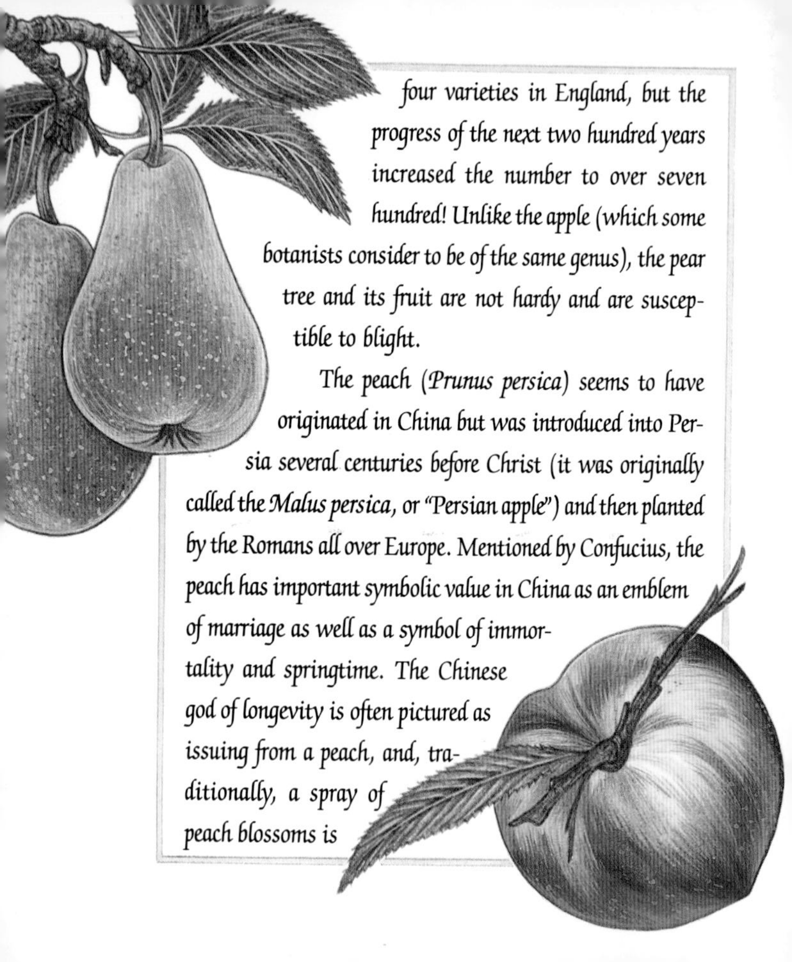

four varieties in England, but the progress of the next two hundred years increased the number to over seven hundred! Unlike the apple (which some botanists consider to be of the same genus), the pear tree and its fruit are not hardy and are susceptible to blight.

The peach (*Prunus persica*) seems to have originated in China but was introduced into Persia several centuries before Christ (it was originally called the *Malus persica*, or "Persian apple") and then planted by the Romans all over Europe. Mentioned by Confucius, the peach has important symbolic value in China as an emblem of marriage as well as a symbol of immortality and springtime. The Chinese god of longevity is often pictured as issuing from a peach, and, traditionally, a spray of peach blossoms is

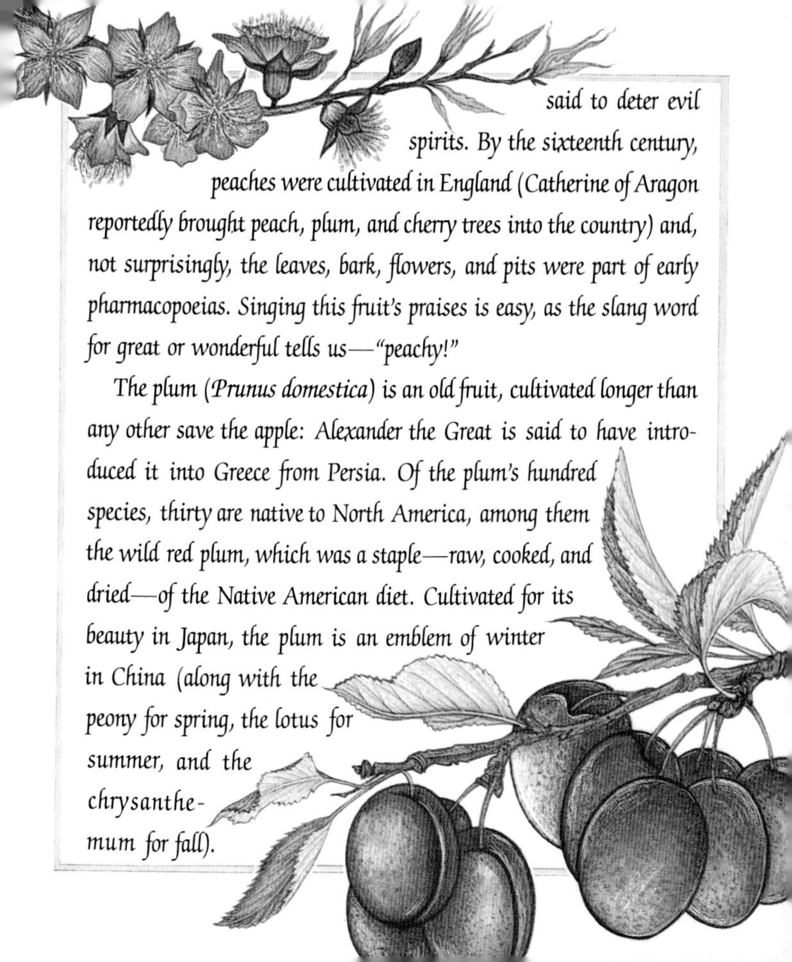

said to deter evil spirits. By the sixteenth century, peaches were cultivated in England (Catherine of Aragon reportedly brought peach, plum, and cherry trees into the country) and, not surprisingly, the leaves, bark, flowers, and pits were part of early pharmacopoeias. Singing this fruit's praises is easy, as the slang word for great or wonderful tells us—"peachy!"

The plum (*Prunus domestica*) is an old fruit, cultivated longer than any other save the apple: Alexander the Great is said to have introduced it into Greece from Persia. Of the plum's hundred species, thirty are native to North America, among them the wild red plum, which was a staple—raw, cooked, and dried—of the Native American diet. Cultivated for its beauty in Japan, the plum is an emblem of winter in China (along with the peony for spring, the lotus for summer, and the chrysanthemum for fall).

WATER LILY *Nymphaea odorata* *Nymphaeaceae*

These fragrant aquatic herbs (whose botanical name reminds us of the water nymphs of Greco-Roman legends) are relatives of the Egyptian lotus, associated with the Nile, the Egyptian "provider of life." The bud and flower of the water lily were the basis of Egyptian architectural motifs. The plant is also sacred in Buddhism, where it symbolizes perfection, since it grows out of the mud but is not defiled by its origins (just as Buddha is considered of this earth but nonetheless above it).

Even though the Native Americans cooked and ground the seeds of the *Nymphaea odorata* into meal, it's not likely that the plants consumed by the Lotus-Eaters in Homer's *Odyssey*—the plants of oblivion and forgetfulness—and tasted by some of Odysseus' followers were water lilies at all. The plant's majestic pads—particularly of those South American species whose pads can measure as much as six or nine feet across—give it a strange, otherworldly beauty.

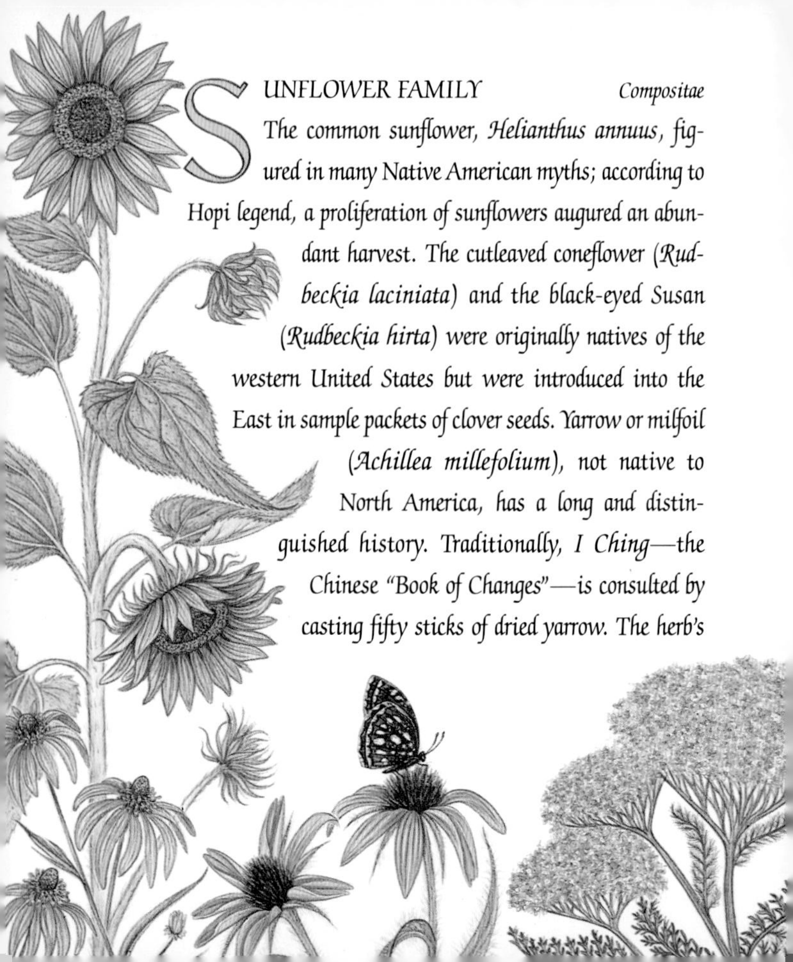

SUNFLOWER FAMILY *Compositae*

The common sunflower, *Helianthus annuus*, figured in many Native American myths; according to Hopi legend, a proliferation of sunflowers augured an abundant harvest. The cutleaved coneflower (*Rudbeckia laciniata*) and the black-eyed Susan (*Rudbeckia hirta*) were originally natives of the western United States but were introduced into the East in sample packets of clover seeds. Yarrow or milfoil (*Achillea millefolium*), not native to North America, has a long and distinguished history. Traditionally, *I Ching*—the Chinese "Book of Changes"—is consulted by casting fifty sticks of dried yarrow. The herb's association with divination has a European origin: a "plant of the Devil," it was called "devil's nettle" or "bad man's plaything" and was used in spells and to ward off evil spirits.

The oxeye daisy (*Chrysanthemum leucanthemum*) was probably introduced to North America unknowingly by the Puritans as they threw out their packing straw. The Romans had dedicated this flower to Diana, goddess of the moon and the hunt, while Margaret of Anjou took it as her emblem during the Wars of the Roses, from which comes one of the flower's common names, marguerite. Girls pluck its petals to see if "he loves me" or "he loves me not," while another legend says that dreaming of daisies in spring or summer portends good luck (and bad luck in the other seasons). Chicory (*Cichorium intybus*) was imported here by the colonists for salads and as a substitute for coffee.

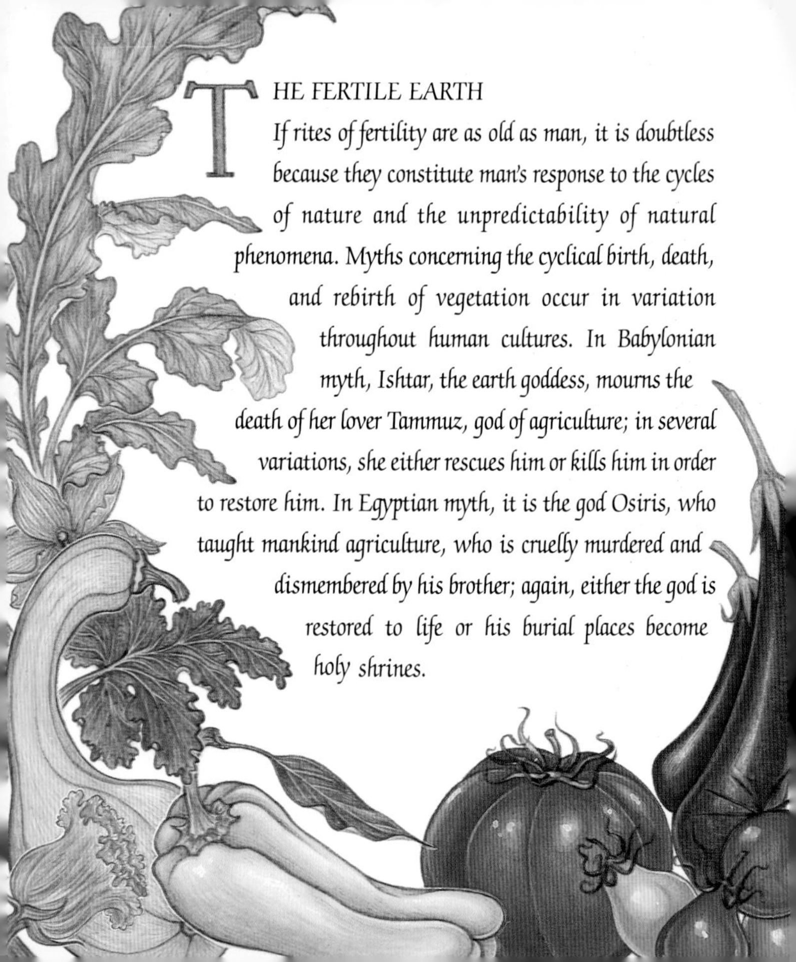

THE FERTILE EARTH

If rites of fertility are as old as man, it is doubtless because they constitute man's response to the cycles of nature and the unpredictability of natural phenomena. Myths concerning the cyclical birth, death, and rebirth of vegetation occur in variation throughout human cultures. In Babylonian myth, Ishtar, the earth goddess, mourns the death of her lover Tammuz, god of agriculture; in several variations, she either rescues him or kills him in order to restore him. In Egyptian myth, it is the god Osiris, who taught mankind agriculture, who is cruelly murdered and dismembered by his brother; again, either the god is restored to life or his burial places become holy shrines.

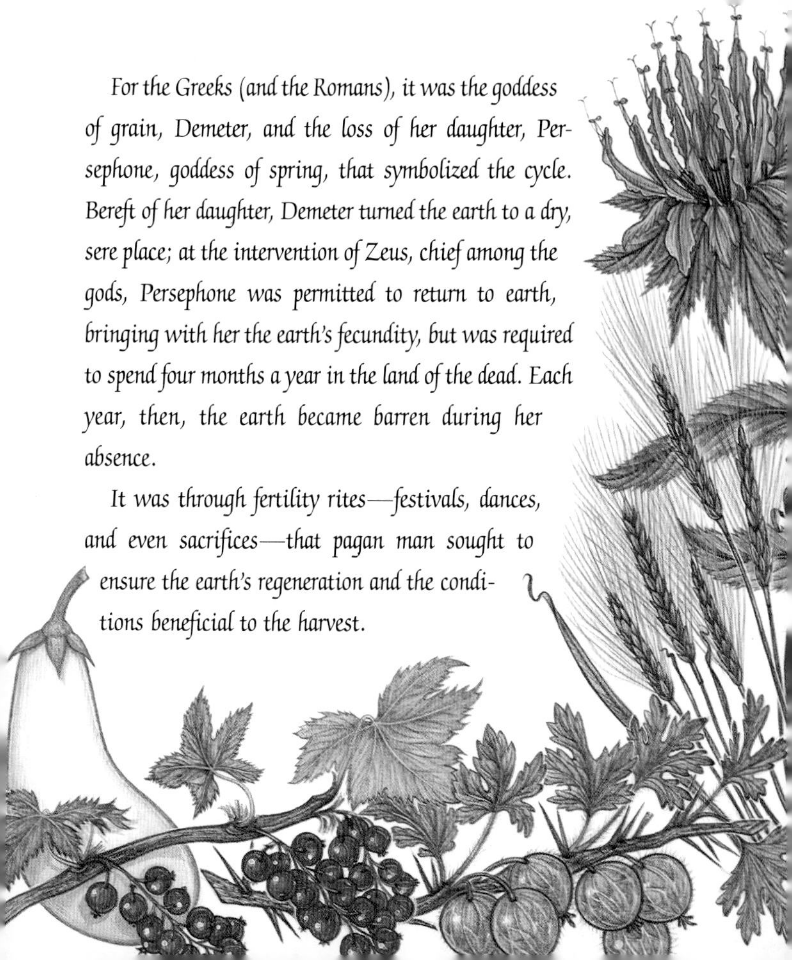

For the Greeks (and the Romans), it was the goddess of grain, Demeter, and the loss of her daughter, Persephone, goddess of spring, that symbolized the cycle. Bereft of her daughter, Demeter turned the earth to a dry, sere place; at the intervention of Zeus, chief among the gods, Persephone was permitted to return to earth, bringing with her the earth's fecundity, but was required to spend four months a year in the land of the dead. Each year, then, the earth became barren during her absence.

It was through fertility rites—festivals, dances, and even sacrifices—that pagan man sought to ensure the earth's regeneration and the conditions beneficial to the harvest.

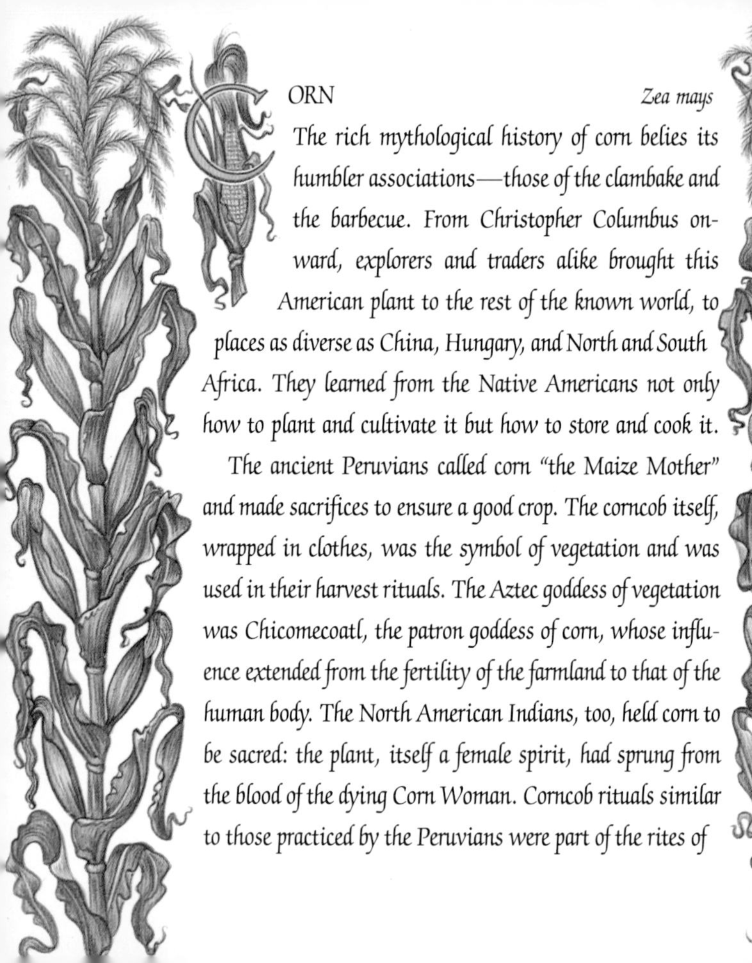

CORN *Zea mays*

The rich mythological history of corn belies its humbler associations—those of the clambake and the barbecue. From Christopher Columbus onward, explorers and traders alike brought this American plant to the rest of the known world, to places as diverse as China, Hungary, and North and South Africa. They learned from the Native Americans not only how to plant and cultivate it but how to store and cook it.

The ancient Peruvians called corn "the Maize Mother" and made sacrifices to ensure a good crop. The corncob itself, wrapped in clothes, was the symbol of vegetation and was used in their harvest rituals. The Aztec goddess of vegetation was Chicomecoatl, the patron goddess of corn, whose influence extended from the fertility of the farmland to that of the human body. The North American Indians, too, held corn to be sacred: the plant, itself a female spirit, had sprung from the blood of the dying Corn Woman. Corncob rituals similar to those practiced by the Peruvians were part of the rites of the Cochiti in New Mexico and the Seminoles in Florida. An Iroquois myth speaks eloquently to the extraordinary centrality of corn in Indian life: the Spirit of the Maize, the Spirit of the Bean, and the Spirit of the Squash were three beautiful sisters who lived together in harmony. The reality of farming is mirrored in the myth: when all three are planted together, the squash prevents the growth of weeds, while the tall cornstalk provides a perch for the beans to grow on. The three are frequently eaten together as well, a traditional practice that, modern nutritionists have pointed out, renders corn nutritionally complete.

While corn itself was a symbol of sacred unity, its many hues were symbolic of its divine multiplicity. According to Zuni legend, when seven maidens were sacrificed, each became a different color of corn. The maidens were resurrected as goddesses who intervened in human life.

MELONS

Cucurbitaceae

Summer would lose a great deal of its sweetness without these kissing cousins of the gourd, pumpkin, and squash. The origin of the muskmelon (*Cucumis melo*) is somewhat of a mystery. Unknown to the Greeks, it is first mentioned in Pliny. In the early part of the seventeenth century, herbalist John Parkinson reported that several varieties were being grown in England but that the best were still being imported from Spain and France. Melons were a part of

the summer luxury that Andrew Marvell described in his 1681 poem "The Garden":

What wond'rous life is this I lead!
Ripe apples drop about my head;
The luscious clusters of the vine
Upon my mouth do crush their wine;
The nectarine and curious peach
Into my hands themselves do reach;
Stumbling on melons, as I pass,
Insnared with flowers, I fall on grass.

While there is indeed a melon called the cantaloupe in Europe, this is a hard-shelled or rock melon; what is called a cantaloupe in the United States is a variety of muskmelon.

GRAPES *Vitaceae*

From the beginning of time, the grape and the vine upon which it grows have been important and powerful symbols. For the ancient Greeks, Dionysus was the god not only of wine but also of fertility, creativity, and the arts. It was said that he wandered through the world teaching men to cultivate the vine, followed by satyrs and nymphs who engaged in raucous and drunken revelry. For the Romans, it was Bacchus who protected the vine. In the Old Testament, the vine remained a powerful symbol, this time of the relationship between God and Man: God is the keeper of the vineyard, while the children of God are likened to the vines He protects. In the New Testament, too, the grapevine is an important emblem of Christ, who declares, "I am the true vine . . ." (John 15:1).

From the time of Dioscorides, wine was considered medicinal, although from the beginning the amount of wine that was actually beneficial remained a matter of debate. Centuries later, John Milton summed the problem

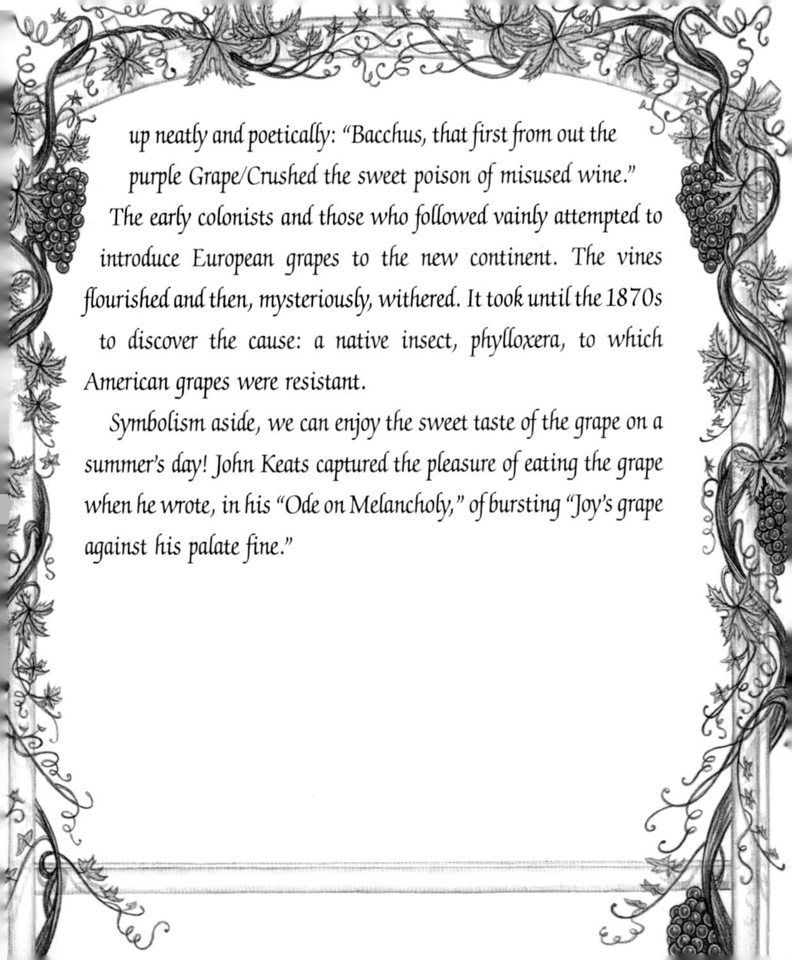

up neatly and poetically: "Bacchus, that first from out the purple Grape/Crushed the sweet poison of misused wine."

The early colonists and those who followed vainly attempted to introduce European grapes to the new continent. The vines flourished and then, mysteriously, withered. It took until the 1870s to discover the cause: a native insect, phylloxera, to which American grapes were resistant.

Symbolism aside, we can enjoy the sweet taste of the grape on a summer's day! John Keats captured the pleasure of eating the grape when he wrote, in his "Ode on Melancholy," of bursting "Joy's grape against his palate fine."

QUEEN ANNE'S LACE *Daucus carota* *Umbelliferae*

The exquisite delicacy of this flower is easy to overlook because, paradoxically, we see it everywhere and in the most unglamorous places—from roadsides to railroad trestles and even on landfills. Yet this pretty wildflower isn't a native of North America at all, but escaped from the gardens of the early colonists, who brought it here as a cultivated plant. (It was grown as an ornamental as late as the nineteenth century in England.) In the sixteenth century, this plant, the wild carrot, was still exotic: it is said that at Queen Elizabeth I's summer court, the ladies wore its leaves, rather than plumes, in their hair. The intricate structure of Queen Anne's lace seems just as marvelous as it did centuries ago: more than sixty thousand petals adorn the plant in its full glory. Its shape after flowering yields its common name, "bird's nest."

The legend of how the plant got its name has many variations, but our favorite is the one in which Queen Anne challenged the ladies of her court to produce lacework as lavish as the pattern of the flower. The Queen herself won the contest, and hence the flower's name.

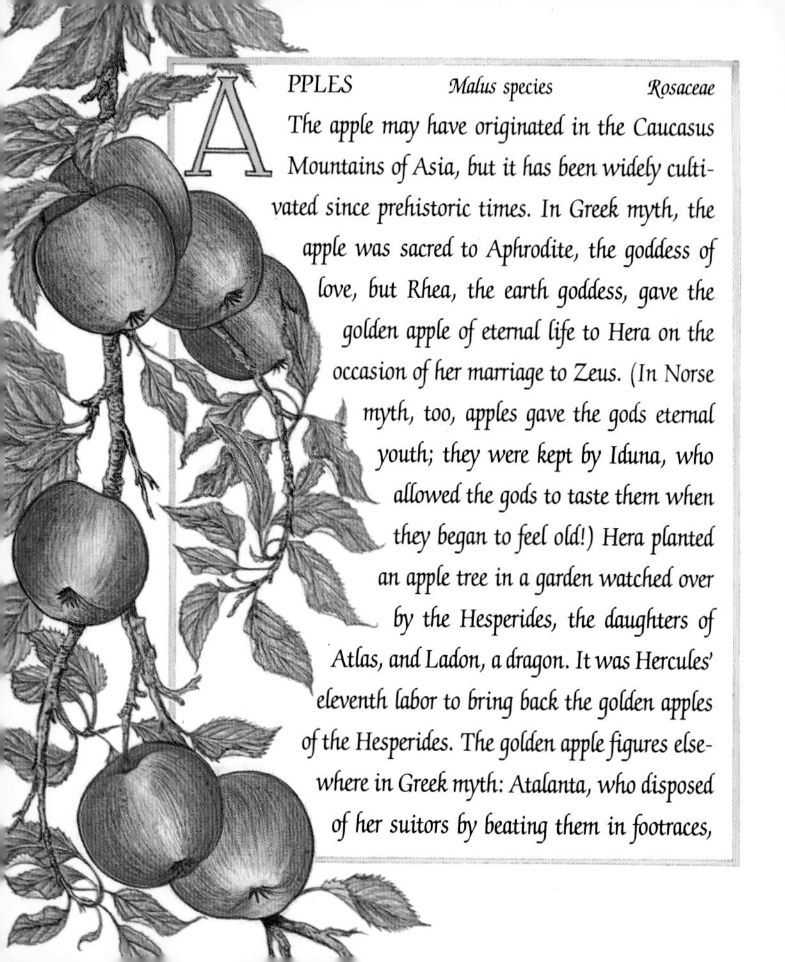

APPLES *Malus species* *Rosaceae*

The apple may have originated in the Caucasus Mountains of Asia, but it has been widely cultivated since prehistoric times. In Greek myth, the apple was sacred to Aphrodite, the goddess of love, but Rhea, the earth goddess, gave the golden apple of eternal life to Hera on the occasion of her marriage to Zeus. (In Norse myth, too, apples gave the gods eternal youth; they were kept by Iduna, who allowed the gods to taste them when they began to feel old!) Hera planted an apple tree in a garden watched over by the Hesperides, the daughters of Atlas, and Ladon, a dragon. It was Hercules' eleventh labor to bring back the golden apples of the Hesperides. The golden apple figures elsewhere in Greek myth: Atalanta, who disposed of her suitors by beating them in footraces, was finally vanquished when she paused to pick up the golden apples thrown by her challenger, Milanion; and the golden apple of discord begat a beauty contest among three goddesses that resulted in the Trojan War. In Roman times, the apple was a love token, and was used in marriage ceremonies.

While the fruit of the forbidden Tree of Knowledge in the Garden of Eden is never identified in the Bible, it is traditionally represented as an apple, doubtless because the Latin words for "apple" (*malum*) and "evil" (*malus*) are so similar. It is an apple that Albrecht Dürer's Eve holds in her hand, and, in John Milton's *Paradise Lost*, Satan declares, "Him by fraud I have seduc'd/From his Creator . . . with an apple." The ancient power of the fruit to lead immortals and mortals astray is doubtless a testament to its beauty and taste. The apple has also long been associated with magic and magical powers. In Celtic myth, the island of the dead where King Arthur and other heroes were taken so that they might someday return to life was called Avalon, from the Welsh word for "apple" (*aval*). It was Avalon's apple trees that gave it its magical power. The Druids used both the apple and the tree for divination; bobbing for apples is a vestige of that tradition.

The apple was cultivated by the Hittites, the Egyptians, and the Romans, who planted the first apple trees in England. The early Spanish explorers tried to introduce the apple to the New World, but the subtropical climates in which they landed were hostile to its growth. It was one William Blackstone, a minister and a student of apple growing, who introduced one new variety in England (the yellow sweeting) and who planted an orchard on Beacon Hill in the Massachusetts Colony in 1623. He stayed on after the first wave of colonists left, and when John Winthrop and the new band of settlers returned in 1630, Blackstone's trees had already begun to bear fruit. Unfortunately, Blackstone didn't fare as well as his orchards: Winthrop banished him to Rhode Island, where he had to start anew. By 1646, the orchards had become sufficiently important to the Colony of

Massachusetts that a law was passed to punish those who robbed one of its fruit.

As the pioneers moved west, so did the apple orchards. Some of these settlers were, of course, aided in their efforts by none other than Johnny Appleseed, more formally known as John Chapman. Now part of American folklore, Chapman appears to have been a religious eccentric (he was first spotted half-naked in a snowstorm, in an ecstatic state) who, for more than forty years, taught others how to cultivate the apple tree.

By 1900, some thousand varieties of apples were being sold in markets across the United States—for apple lovers, a long-past golden age indeed!

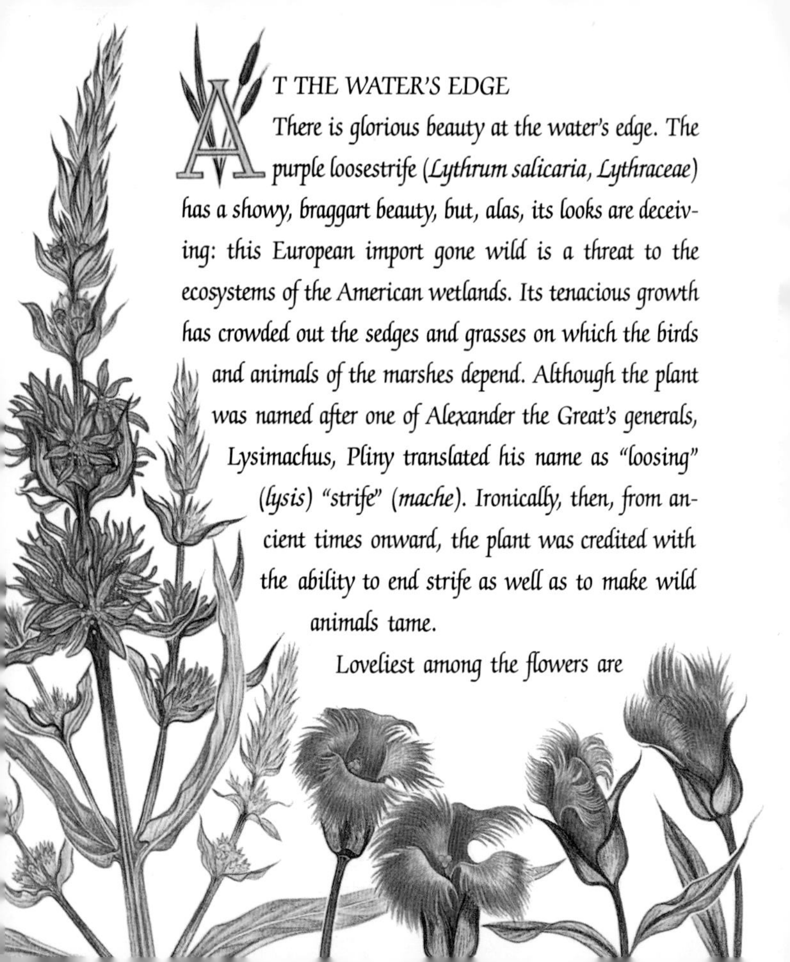

AT THE WATER'S EDGE

There is glorious beauty at the water's edge. The purple loosestrife (*Lythrum salicaria, Lythraceae*) has a showy, braggart beauty, but, alas, its looks are deceiving: this European import gone wild is a threat to the ecosystems of the American wetlands. Its tenacious growth has crowded out the sedges and grasses on which the birds and animals of the marshes depend. Although the plant was named after one of Alexander the Great's generals, Lysimachus, Pliny translated his name as "loosing" (*lysis*) "strife" (*mache*). Ironically, then, from ancient times onward, the plant was credited with the ability to end strife as well as to make wild animals tame.

Loveliest among the flowers are

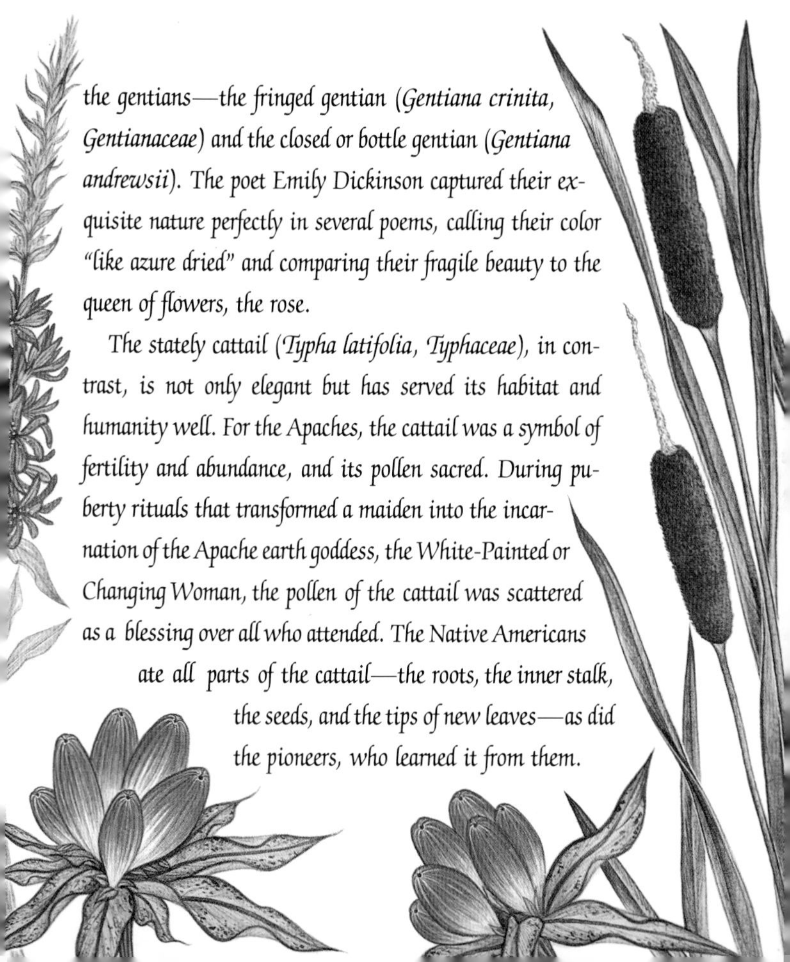

the gentians—the fringed gentian (*Gentiana crinita, Gentianaceae*) and the closed or bottle gentian (*Gentiana andrewsii*). The poet Emily Dickinson captured their exquisite nature perfectly in several poems, calling their color "like azure dried" and comparing their fragile beauty to the queen of flowers, the rose.

The stately cattail (*Typha latifolia, Typhaceae*), in contrast, is not only elegant but has served its habitat and humanity well. For the Apaches, the cattail was a symbol of fertility and abundance, and its pollen sacred. During puberty rituals that transformed a maiden into the incarnation of the Apache earth goddess, the White-Painted or Changing Woman, the pollen of the cattail was scattered as a blessing over all who attended. The Native Americans ate all parts of the cattail—the roots, the inner stalk, the seeds, and the tips of new leaves—as did the pioneers, who learned it from them.

TO AUTUMN

Season of mists and mellow fruitfulness,
   Close bosom-friend of the maturing sun,
Conspiring with him how to load and bless
   With fruit the vines that round the thatch-eves run;
To bend with apples the moss'd cottage-trees,
   And fill all fruit with ripeness to the core;
      To swell the gourd, and plump the hazel shells
   With a sweet kernel; to set budding more,
And still more, later flowers for the bees,
Until they think warm days will never cease,
      For Summer has o'er-brimmed their clammy cells.

Who hath not seen thee oft amid thy store?
   Sometimes whoever seeks abroad may find
Thee sitting careless on a granary floor,
   Thy hair soft-lifted by the winnowing wind;
Or on a half-reaped furrow sound asleep,

Drowsed with the fume of poppies, while thy hook
Spares the next swath and all its twined flowers;
And sometimes like a gleaner thou dost keep
Steady thy laden head across a brook;
Or by a cider-press, with patient look,
Thou watchest the last oozings hours by hours.

Where are the songs of Spring? Ay, where are they?
Think not of them, thou hast thy music too,—
While barred clouds bloom the soft-dying day,
And touch the stubble-plains with rosy hue:
Then in a wailful choir the small gnats mourn
Among the river sallows, borne aloft
Or sinking as the light wind lives or dies;
And full-grown lambs loud bleat from hilly bourn;
Hedge-crickets sing; and now with treble soft
The red-breast whistles from a garden-croft;
And gathering swallows twitter in the skies.

—John Keats

PUMPKINS, SQUASH, AND GOURDS *Cucurbitaceae* These plants have been cultivated for so long—since prehistoric times—that little is known of their wild state and origin. The word "pumpkin" (*Cucubita pepo*) is a transformation of the Latin word *peponem*, which in turn became *pompion* in Old French, "pompion" or "pumpion" in English, and, finally, "pumpkin." (In 1828, Washington Irving still called it a "pompion.") The Apaches, who harvested pumpkins, held ceremonies to ensure a good crop, first barring newlyweds and pregnant women from the pump-

kin patch so that they did not "rob" the harvest of the "fruit of life." A young boy was then sent to scatter juniper berries over the earth in all directions; it was thought that the plot would yield pumpkins wherever the berries had been scattered. Pumpkin was, of course, served at the first Thanksgiving feast in 1621.

Our word "squash," on the other hand, is a shortened version of the Narraganset Indian word *asquutasquash*, meaning "raw" or "uncooked." Gourds, those *Cucurbitaceae* that have hard, durable shells, have long been used as utensils, as well as musical instruments.

AUTUMN LEAVES

Trees have long figured in man's life and in his myths; no doubt the symbolic history of trees is as old as they are. The Romans identified certain trees with gods: the strong beech with Hercules and Diana, goddess of the hunt; and the ash, from which spears and poles were fashioned, with Mars, god of war. In Nordic myth, the great ash was the tree of the cosmos, Yggdrasil, which sent its roots into the earth's core (where the underworld was) and whose branches reached up through the heavens. The world tree—though not always an ash—appears in many North American Indian myths as well; in Iroquois myth, the fertility goddess falls to earth from the roots of the tree of the heavens.

Just as the budding of the trees in spring is the first sign of the earth's rebirth, so, too, the vivid hues of autumn signal winter's approach. It is a time of beauty, as Henry David Thoreau's words remind us:

> Already, by the first of September, I had seen two or three small maples turned scarlet across the pond, beneath where the white stems of three aspens diverged, at the point of a

promontory, next the water. Ah, many a tale their color told! And gradually from week to week the character of each tree came out, and it admired itself reflected in the smooth mirror of the lake. Each morning the manager of this gallery substituted some new picture, distinguished by more brilliant or harmonious coloring, for the old upon the walls.

The "gallery" on these pages (to borrow from Thoreau) comprises but a tiny piece of what nature has to offer. The flowering dogwood (*Cornus florida, Cornaceae*) is most familiar when it is in flower, but in fall its leaves turn a bright, vibrant red. The leaf of the ginkgo (*Ginkgo biloba, Ginkgoaceae*) is by far the oldest specimen on these pages. The Ginkgo, or maidenhair, tree is a living fossil, the only remaining and unchanged species of a large class of trees, the gymnosperms, from earth's Triassic period. In China and Japan, where the ginkgo is native, it is considered sacred and planted near temples. It is familiar to urban dwellers, though, not because of its extraordinary history but for its distinctive hardiness: it thrives amid dirt, pollution, and even smog!

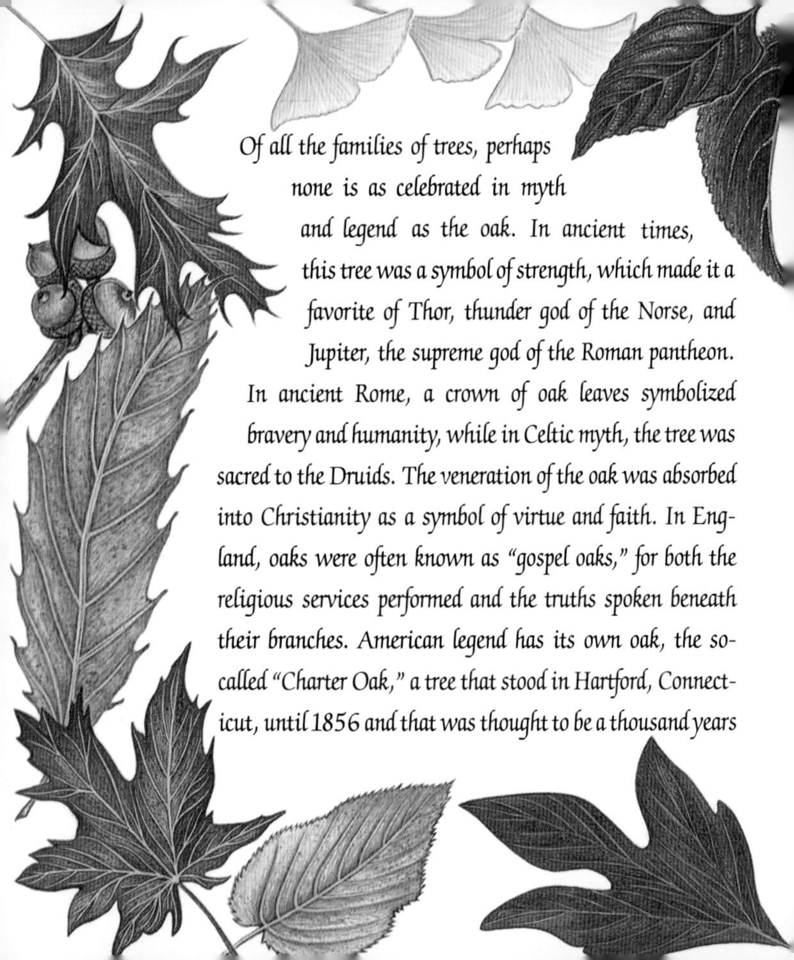

Of all the families of trees, perhaps none is as celebrated in myth and legend as the oak. In ancient times, this tree was a symbol of strength, which made it a favorite of Thor, thunder god of the Norse, and Jupiter, the supreme god of the Roman pantheon. In ancient Rome, a crown of oak leaves symbolized bravery and humanity, while in Celtic myth, the tree was sacred to the Druids. The veneration of the oak was absorbed into Christianity as a symbol of virtue and faith. In England, oaks were often known as "gospel oaks," for both the religious services performed and the truths spoken beneath their branches. American legend has its own oak, the so-called "Charter Oak," a tree that stood in Hartford, Connecticut, until 1856 and that was thought to be a thousand years

old. Tradition has it that when, in 1687, the governor general of New England, Sir Edmund Andros, demanded that the colonists surrender the charter of Connecticut, it was hidden in the tree. Shown here are the leaves of the pin oak (*Quercus palustris, Fagaceae*) and, on the following page, the white oak (*Quercus alba, Fagaceae*). The stately white oak was treasured by the Native Americans, who ground the acorns into flour, as did the pioneers, following the Indians' example. The tree is also known as stave oak, since it is prized for use in whiskey barrels.

It's hard to look at the leaf of the American chestnut (*Castanea dentata, Fagaceae*) without thinking of Henry Wadsworth Longfellow's village smithy standing under the tree's spread! On the other hand, the brilliant leaf of the sugar or rock maple (*Acer saccharum*, *Aceraceae*) brings something else to mind: the sweet velvet of maple syrup. In the late eighteenth and early nineteenth centuries the sugar that maple syrup yielded was the choice of antislavery advocates, who called it the sugar of "free men," as opposed to the cane sugar harvested by slaves on plantations. It is the sweet or black birch (*Betula lenta*, *Betulaceae*) that yields oil of wintergreen. Is this, we wonder, the kind of birch from which Robert Frost's boy swung in his famous poem "Birches"?

The sixteenth and seventeenth centuries marked both the discovery and colonization of America *and* the great age of herbal medicine in England and Europe. The American continent soon became a source for new curatives, among them the native sassafras (*Sassafras albidum*, *Lauraceae*), which was exported by the shipload. Sassafras was probably discovered as early as the 1520s by Spanish explorers, who touted it as a cure for rheumatism and syphilis. Sassafras bark and root were sold well into the nineteenth century as flavorings and curatives by the Shaker community and others.

The so-called "Yellow poplar" isn't a poplar at all but rather a member of the magnolia family, the beautiful and showy tulip tree (*Liriodendron tulipifera, Magnoliaceae*). The wood is a favorite of cabinetmakers. The leaf of the shining sumac (*Rhus copallina, Anacardiaceae*) completes the "gallery" and does indeed shine in both its summer greenery and its autumnal red.

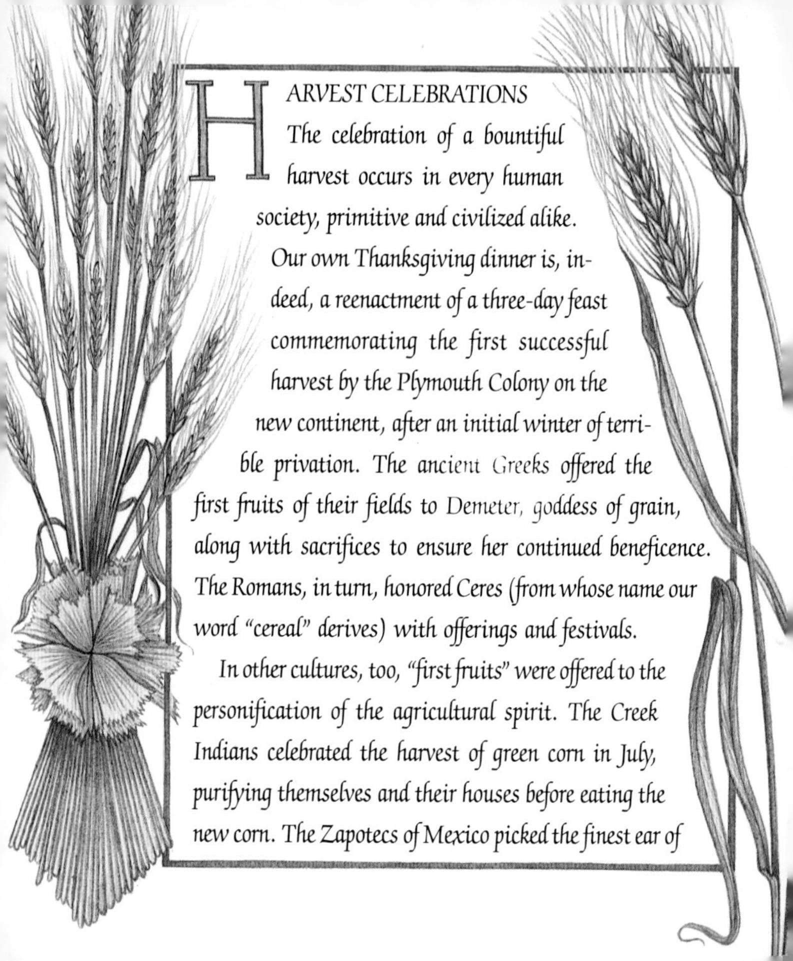

HARVEST CELEBRATIONS

The celebration of a bountiful harvest occurs in every human society, primitive and civilized alike. Our own Thanksgiving dinner is, indeed, a reenactment of a three-day feast commemorating the first successful harvest by the Plymouth Colony on the new continent, after an initial winter of terrible privation. The ancient Greeks offered the first fruits of their fields to Demeter, goddess of grain, along with sacrifices to ensure her continued beneficence. The Romans, in turn, honored Ceres (from whose name our word "cereal" derives) with offerings and festivals.

In other cultures, too, "first fruits" were offered to the personification of the agricultural spirit. The Creek Indians celebrated the harvest of green corn in July, purifying themselves and their houses before eating the new corn. The Zapotecs of Mexico picked the finest ear of corn, adorned it with clothes and jewelry, and honored it with songs, dances, and sacrifices. Even among people who did not farm, rituals of thanksgiving existed in similar forms: the Salish and Tinneh Indians thanked the plants that yielded fruits and berries before eating them, while fishing tribes gave thanks to the returning salmon.

Other traditions emphasize the *last* sheaf of grain taken from the fields, in which the agricultural spirit is thought to reside. Although there are many variations on this custom (observed in places as diverse as Denmark, Poland, Brittany, Scotland, Ireland, England, and Germany), often the last sheaf taken from the field is toasted or serenaded; at other times, it is fashioned into a female figure (called the "Harvest Queen" or "Corn Baby"), which is then feted by the reapers with song and dance, or is ceremoniously carried home by the reapers and

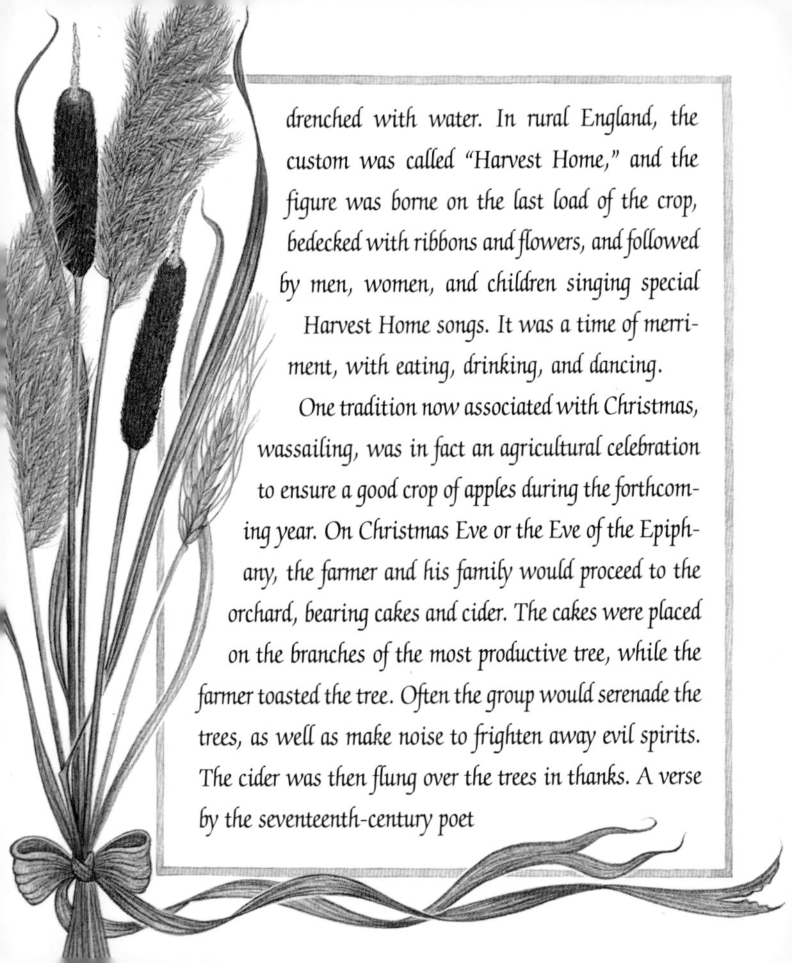

drenched with water. In rural England, the custom was called "Harvest Home," and the figure was borne on the last load of the crop, bedecked with ribbons and flowers, and followed by men, women, and children singing special Harvest Home songs. It was a time of merriment, with eating, drinking, and dancing.

One tradition now associated with Christmas, wassailing, was in fact an agricultural celebration to ensure a good crop of apples during the forthcoming year. On Christmas Eve or the Eve of the Epiphany, the farmer and his family would proceed to the orchard, bearing cakes and cider. The cakes were placed on the branches of the most productive tree, while the farmer toasted the tree. Often the group would serenade the trees, as well as make noise to frighten away evil spirits. The cider was then flung over the trees in thanks. A verse by the seventeenth-century poet

Robert Herrick makes it clear that the tradition was widespread, and not necessarily limited to apples:

Wassaile the Trees, that they may beare
You many a Plum, and many a Peare:
For more or lesse fruits they will bring,
As you doe give them Wassailing.

Who loves a garden still his Eden keeps,
Perennial pleasures plants, and wholesome harvests reaps.

—AMOS BRONSON ALCOTT

In addition to the flowers, fruits, and vegetables identified in the text, others are pictured in the borders and double-page spreads. What follows is a key, reading left to right in all cases but one:

hepatica, monkshood, tall buttercup: full-page painting, "Buttercup Family"
larkspur, marsh marigold: border, "Buttercup Family"
peach blossom: top right, fourth page, "Rose Family"
common reed: border of "Water Lily"
arugula, crook neck squash and blossom, Italian pepper, tomato, purple eggplant, white eggplant, currants, gooseberries, wheat, bergamot: "The Fertile Earth"
McIntosh, Newtown pippin, Granny Smith, Jonathan: "Apples"

bittersweet: border, John Keats's "To Autumn"

watermelon, Crenshaw, cantaloupe, casaba, honeydew: "Melons"

pumpkin, Turk's turban squash, crookneck squash, loofah gourd: "Pumpkins"

red maple leaf, bigtooth Aspen leaf, American beech leaf, American elm leaf, white ash leaflet: border "Autumn Leaves"

wheat: "Harvest Celebrations"

The endpapers of this book depict fallen leaves on a forest floor. Pictured on the jacket are dwarf goldenrod, apples, black oak leaves, Indian corn, pumpkin, silver maple leaf, purple grapes, dwarf sunflower, common reeds, and cattails.